THE ULTIMATE
SPACE
FACTS

FOR KIDS, TEENS, & ADULTS

Blast Off on a Journey Through the Cosmos and Discover Mind-Blowing Facts, Records, Mysteries, Inventions, and Wonders of the Universe!

Book 4 of Eleven Worlds to Explore
Ethereal Ray

Table of Contents

Hi there! I'm **Luna**, and this is my spacefaring cat, **Star**. We're your guides to this awesome collection of space facts. We can't wait for you to discover all the amazing things about the universe! If you enjoy this cosmic adventure, maybe you could ask a grown-up to leave a review on **Amazon**? **Reviews** help us explore new galaxies and write even more exciting space books for kids like you!

Review us on **Amazon US!**

or any Amazon site where you purchased this book!

> *"The cosmos is all that is or ever was or ever will be."*

- Carl Sagan

Get ready to blast off on an incredible journey through the wonders of space! "**The Ultimate Space Facts for Kids, Teens, & Adults:** Blast Off on a Journey Through the Cosmos and Discover Mind-Blowing Facts, Records, Mysteries, Inventions, and Wonders of the Universe!" is packed with amazing information, stunning visuals, and fun activities designed to ignite your curiosity and expand your knowledge of the universe.

This book is more than just a collection of facts—it's a passport to the cosmos:

- **Ignite your imagination:** Explore the planets of our solar system, journey to distant stars, and unravel the mysteries of black holes. Each chapter will spark your curiosity and transport you to the far reaches of space.

- **Fuel your mind:** Discover how rockets work, learn about the challenges of living in space, and delve into the fascinating world of space inventions. This book is an intellectual launchpad for curious minds of all ages.

- **Expand your horizons:** From the smallest asteroid to the largest galaxy, uncover the secrets of the universe and gain a deeper understanding of our place in the cosmos.

- **Bond through discovery:** Share this space adventure with family and friends. Explore the wonders of the universe together, create your own solar system models, and embark on a journey of learning and discovery.

Embark on an exploration of stars, planets, galaxies, and beyond. With each page you turn, you'll uncover new wonders of the universe and deepen your appreciation for the vastness of space.

So, put on your spacesuit, buckle up, and prepare for liftoff! Your cosmic adventure awaits !

Chapter 1:
Welcome to the Universe!

Have you ever looked up at the night sky and wondered, "What's out there?"

Get ready to blast off on an incredible journey as we explore the amazing universe!

In this chapter, we'll uncover mind-blowing facts about **space**, from the sheer size of the cosmos to the mind-boggling number of stars.

We'll discover what makes our planet special and learn about the incredible objects that fill the vastness of space.

So, put on your space helmets, fire up your imaginations, and prepare for liftoff! It's time to embark on a cosmic adventure that will take you to the farthest reaches of the universe!

What is space?

Space is everything beyond our planet Earth. It's where our Sun, Moon, and all those stars hang out. It's a gigantic place with no air to breathe, and you'd float around like a feather if you were there! In space, you'll find incredible things like:

- **Stars:** Giant balls of burning gas that give off light and heat, just like our Sun. Some stars are even bigger than our Sun – can you imagine that?

- **Planets:** Worlds that orbit stars. Some are big and gassy like Jupiter, with swirling storms and colorful clouds. Others are rocky like our Earth, with mountains, valleys, and maybe even oceans!

- **Galaxies:** Massive collections of stars, gas, and dust. Our galaxy is called the Milky Way, and it's shaped like a giant pinwheel! There are billions of galaxies in the universe, each containing billions of stars. Whoa!

- **Nebulae:** Giant clouds of gas and dust where stars are born. They come in incredible colors and shapes, like the Eagle Nebula, which looks like a giant bird with outstretched wings.

- **Black holes:** Mysterious objects with such strong gravity that nothing, not even light, can escape them! Imagine a giant vacuum cleaner in space, sucking everything in!

But wait, there's more!

Space is also filled with other fascinating things:

- **Asteroids:** Giant rocks that zoom around the Sun. Some are as small as a car, while others are as big as a city!

- **Comets:** Like giant snowballs made of ice and dust. When they get close to the Sun, they heat up and leavea beautiful trail of light.

- **Satellites:** Machines launched from Earth that orbit our planet. They help us do all sorts of things, like make phone calls, watch TV, and even see where we are on a map!

How big is space?

Space is so mind-bogglingly big that it's hard to even imagine! Our Earth seems like a giant place to us, but compared to space, it's like a tiny speck of dust. Here's a fun way to think about it:

- If the Sun were the size of a basketball, Earth would be about the size of the head of a pin! That's super tiny!
- And the Milky Way galaxy? Well, if the Sun were that basketball, the Milky Way would be as big as the entire continent of North America!

To measure the enormous distances in space, we use something called **light-years**. A light-year is the distance that light travels in one year. Light is the fastest thing in the universe, but it still takes a long time to get anywhere in space. And guess what? It would take millions of years to travel to the nearest star outside our solar system, even if we could travel at the speed of light!

What is the universe?

The universe is everything! It's all of space and everything in it. That includes you, me, our planet Earth, the Sun, the Moon, all those twinkling stars, and even things we haven't discovered yet.

Scientists believe that the universe began with a massive explosion called the **Big Bang**, about 13.8 billion years ago. Since then, the universe has been expanding, getting bigger and bigger all the time! Imagine a balloon being blown up – that's kind of like what the universe is doing.

Isn't space amazing?

Just thinking about how vast and mysterious space is can make you feel a sense of wonder. Every time we look up at the stars, we're seeing light that has traveled for millions and billions of years to reach us. And who knows what incredible secrets are still waiting to be discovered out there in the darkness? It's like a giant puzzle, and we're just starting to put the pieces together!

Fun facts about space:

- There are more stars in the universe than there are grains of sand on all the beaches on Earth! Can you believe that?

- Space is completely silent because there's no air for sound to travel through. Imagine how peaceful that would be!

- Nobody knows exactly how big the universe is! It might even go on forever.

A Cosmic Perspective

Exploring space has not only taught us about the vastness of the universe but also given us a new appreciation for our own planet. When astronauts look back at Earth from space, they see a beautiful, fragile sphere floating in the darkness. This "cosmic perspective" reminds us that Earth is our only home, and we need to protect it.

- **One Big Family:** From space, you can't see any borders between countries. Earth looks like one big, interconnected planet. This reminds us that we are all part of one global family, and we need to work together to solve problems like climate change and pollution.

- **A Pale Blue Dot:** The famous astronomer Carl Sagan once called Earth a "pale blue dot" when he saw a picture of it taken from billions of miles away by the Voyager 1 spacecraft. This image highlights how small and precious our planet is in the vastness of space.

How Astronomers Explore the Universe

Astronomers are like space detectives, using incredible tools and techniques to unravel the mysteries of the universe. Here are some of their most important tools:

- **Telescopes:** These are like giant eyes that allow us to see distant objects in space. There are different types of telescopes, from optical telescopes that collect visible light to radio telescopes that detect radio waves from space.

- **Spectroscopy:** This technique allows astronomers to study the light from stars and galaxies, revealing their composition, temperature, and movement. It's like taking a "fingerprint" of a star!

- **Spacecraft:** Robotic spacecraft, like the Voyager probes and Mars rovers, travel to other planets and moons, collecting data and sending it back to Earth.

Are you ready to explore more of this amazing universe? Let's continue our journey in the next chapter!

Chapter 2:
Our Solar System
- A Neighbourhood Tour

Get ready to blast off on a tour of our cosmic neighborhood! In this chapter, we'll explore the amazing planets, moons, and other celestial objects that make up our solar system.

From the scorching heat of Mercury to the icy rings of Saturn, we'll uncover fascinating facts and incredible discoveries about these amazing worlds.

So, buckle up and join Luna and Star as we embark on an unforgettable journey through our solar system!

Imagine a giant ball of burning gas, so massive that you could fit over one million Earths inside it! That's our Sun, the star at the center of our solar system. It's like a giant power plant, giving off the light and heat that makes life possible on Earth.

- **What is the Sun made of?** The Sun is mostly made of hydrogen and helium gas. Deep inside the Sun, these gases are squeezed together so tightly that they get super hot and create energy. It's like a giant pressure cooker!

- **How hot is the Sun?** The surface of the Sun is about 10,000 degrees Fahrenheit! That's hot enough to melt anything! But the core of the Sun is even hotter, reaching millions of degrees. If you could get close enough, you'd be instantly vaporized!

- **Layers of the Sun:** The Sun is like an onion, with different layers.

  o The **core** is where the energy is made.
  o Then there's the **radiative zone**, where energy travels slowly outwards.
  o Next is the **convective zone**, where hot gas rises and cool gas sinks, like a boiling pot of water.
  o Finally, there's the **photosphere**, which is the surface of the Sun that we see.

- **Solar flares and sunspots:** Sometimes, the Sun has explosions called **solar flares** that burst out with incredible energy. These flares can disrupt our satellites and even cause power outages on Earth. And sometimes, there are dark spots on the Sun called **sunspots**. These are cooler areas caused by the Sun's magnetic field.

Fun facts about the Sun:

- It takes about 8 minutes for sunlight to reach Earth. So, the sunlight you see now actually left the Sun 8 minutes ago!

- The Sun is so powerful that it creates a **solar wind**, a stream of charged particles that flows out into space. This solar wind can create beautiful auroras, or northern and southern lights, when it interacts with Earth's atmosphere.

- The Sun is about halfway through its life. In billions of years, it will eventually run out of fuel and become a **red giant** star, swelling up and swallowing Mercury and

Venus! But don't worry, that won't happen for a very long time.

The Planets - Our Neighbors

Now that we've met the star of our show, let's get to know the planets that orbit it!

Our solar system has eight planets, all orbiting the Sun. They come in different sizes, colors, and compositions. Let's take a closer look at each one, starting with the inner planets!

Inner Planets (The Rocky Planets)

These planets are called "rocky" because they have solid surfaces made of rock and metal. They're also much smaller than the gas giants in the outer solar system. Imagine them as the cozy little houses in our space neighborhood!

Mercury: The Speedy Little Planet

Mercury is the smallest planet in our solar system and the closest to the Sun. It's like the speedy little scooter zipping around the Sun in just 88 Earth days! That means a year on Mercury is shorter than a year on Earth.

- **Super Speedy:** Mercury is the fastest planet, orbiting the Sun at a speed of almost 107,000 miles per hour! That's like traveling from New York to Los Angeles in less than 3 minutes!

- **Extreme Temperatures:** Because it's so close to the Sun, Mercury has extreme temperatures. During the day, it can get as hot as 800 degrees Fahrenheit! But at night, it gets super cold, dropping to -290 degrees Fahrenheit. Brrr! Imagine having to pack for both a sauna and a freezer!

- **Covered in Craters:** Mercury's surface is covered in craters, like a giant space pincushion. These craters were formed by asteroids and comets crashing into the planet long ago. Some of these craters are huge, like the Caloris Basin, which is over 900 miles wide!

- **No Atmosphere:** Mercury has almost no atmosphere, which means there's no air to breathe and no protection from the Sun's harmful rays. If you were to stand on Mercury without a spacesuit, you'd be fried like an egg!

Fun fact: A day on Mercury is longer than a year on Mercury! It takes Mercury 59 Earth days to rotate once on its axis (a day), but only 88 Earth days to orbit the Sun (a year). That's like having your birthday every three months!

Venus: The Hottest Planet

Venus is the second planet from the Sun, and it's the hottest planet in our solar system, even hotter than Mercury! It's like a giant oven in space!

- **Thick Blanket of Clouds:** Venus is covered in thick clouds that trap heat, creating a runaway greenhouse effect. This makes the surface temperature a scorching 900 degrees Fahrenheit! That's hot enough to melt lead!

- **Volcanoes and Mountains:** Venus has a rocky surface with volcanoes, mountains, and vast plains. It's kind of like Earth, but much hotter and with a poisonous atmosphere. Imagine a landscape filled with erupting volcanoes and lava flows!

- **Spinning Backwards:** Venus spins backward compared to most other planets. That means the Sun rises in the west and sets in the east on Venus. How strange is that? It's like watching a movie in reverse!

Fun fact: Venus is the brightest planet in our night sky, sometimes even visible during the day! It's so bright because its thick clouds reflect sunlight very well.

Earth: Our Home Sweet Home

Earth is the third planet from the Sun, and it's the only planet we know of that has life. It's like a giant spaceship carrying all sorts of living things: plants, animals, and, of course, humans!

- **Just the Right Conditions:** Earth has the perfect conditions for life to thrive. It has liquid water, a breathable atmosphere, and a comfortable temperature. It's like Goldilocks found the perfect planet – not too hot, not too cold, but just right!

- **Oceans and Continents:** Earth is mostly covered in water, with vast oceans and continents. It's the only planet in our solar system with liquid water on its surface. These oceans are home to incredible creatures, from tiny plankton to giant whales!

- **The Moon:** Earth has one moon that orbits around it. The Moon controls the tides and lights up our night sky. It's like our own personal nightlight!

Fun fact: Earth is the only planet in our solar system not named after a Roman god or goddess.

Earth's Moon: Our Nearest Neighbor

Our Moon is Earth's only natural satellite. It's a big ball of rock that orbits our planet, and it's been our companion in space for billions of years.

- **Phases of the Moon:** Have you ever noticed how the Moon seems to change shape? Sometimes it's a full circle, sometimes it's a crescent, and sometimes it disappears altogether! These are the phases of the Moon, caused by the changing angles of sunlight reflecting off its surface.

- **Craters and Maria:** The Moon's surface is covered in craters, caused by asteroids and comets crashing into it long ago. It also has dark, flat areas called maria, which are ancient lava flows.

- **Tides:** The Moon's gravity pulls on Earth's oceans, causing the tides to rise and fall. That's why we have high tide and low tide every day!

- **No Atmosphere:** The Moon has no atmosphere, which means there's no air to breathe and no weather. It's also very quiet on the Moon because sound can't travel without air.

Fun fact: The first human to walk on the Moon was Neil Armstrong in 1969. He famously said, "That's one small step for man, one giant leap for mankind."

Mars: The Red Planet

Mars is the fourth planet from the Sun, and it's known as the Red Planet because of its rusty color. Scientists believe that Mars may have once been much warmer and wetter, with oceans and rivers, just like Earth.

- **Rusty Red:** Mars gets its red color from iron oxide, which is the same thing that makes rust red. Imagine a whole planet covered in rust!

- **Giant Volcanoes:** Mars has the largest volcano in the solar system, Olympus Mons. It's three times taller than Mount Everest! Can you imagine climbing that?

- **Two Tiny Moons:** Mars has two small moons, Phobos and Deimos. They're lumpy and potato-shaped, unlike our round Moon.

- **Searching for Life:** Scientists are searching for signs of past or present life on Mars. They've sent rovers to explore the surface and look for clues. Maybe one day, humans will walk on Mars too!

Fun fact: A day on Mars is almost the same length as a day on Earth, but a year on Mars is almost twice as long! That means you'd have to wait a lot longer for your birthday if you lived on Mars.

Mars: Humanity's Second Home?

For centuries, Mars has captured humanity's imagination as a potential second home. With its striking red surface and similarities to Earth, the planet seems like the most viable option for colonizing beyond our world. But how realistic is this dream?

Why Mars?

Mars is often considered the "Goldilocks" planet of the Solar System—not too close, not too far. Its day, or *sol*, is just 24.6 hours, making it the closest in duration to Earth's day. Mars also has seasons, polar ice caps, and evidence of ancient rivers and lakes, suggesting it once had liquid water.

Moreover, the planet's soil contains resources like water locked in ice, which can potentially be extracted and used for drinking, farming, and producing fuel. Its thin atmosphere, primarily composed of carbon dioxide, could be harnessed to grow plants in controlled environments.

Challenges of Living on Mars

However, the path to making Mars habitable is fraught with challenges:

1. **Radiation Exposure:** Mars lacks a magnetic field, meaning its surface is bombarded by cosmic rays and solar radiation. Long-term exposure could harm settlers without proper shielding.

2. **Extreme Temperatures:** The average temperature on Mars is a chilly -60°C (-80°F), dropping as low as -125°C (-195°F) at the poles.

3. **Thin Atmosphere:** With just 1% of Earth's atmospheric pressure, breathing on Mars without a spacesuit is impossible.

4. **Isolation and Mental Health:** Settlers would face years-long isolation from Earth, which could lead to psychological challenges.

The Role of SpaceX

SpaceX, founded by Elon Musk, is leading the charge to make life on Mars a reality. Its **Starship spacecraft**, designed for interplanetary travel, aims to transport cargo and humans to Mars within the next decade. Musk envisions a self-sustaining city on Mars with a population of one million people by 2050.

To achieve this, SpaceX is working on:

- **Reusable Rockets:** Lowering the cost of space travel.

- **In-Situ Resource Utilization (ISRU):** Extracting water from Martian soil and using carbon dioxide to create fuel and breathable oxygen.

- **Habitat Designs:** Pressurized domes or underground bases to shield settlers from radiation and extreme temperatures.

Terraforming Mars: Dream or Reality?

Terraforming—altering Mars' environment to make it Earth-like—is a long-term vision. Scientists propose ideas like releasing greenhouse gases to warm the planet or using giant mirrors to reflect sunlight onto the surface. However, such projects could take centuries, if not millennia.

For now, humanity's first step is establishing a sustainable base on Mars, akin to a "Martian Antarctica." This would serve as a hub for exploration, research, and resource extraction, paving the way for future colonization.

Looking Ahead

While the challenges are immense, the dream of living on Mars pushes the boundaries of science, engineering, and human determination. The pursuit of Mars colonization also inspires innovation that benefits Earth, from renewable energy solutions to advances in robotics and medicine.

Perhaps, within our lifetimes, humanity will take its first giant leap onto the Red Planet.

Outer Planets (The Gas Giants)

Now, let's journey to the outer solar system, where the gas giants reign! These planets are huge and made mostly of gas, so you couldn't stand on their surface like you could on the rocky planets.

Jupiter: The King of the Planets

Jupiter is the largest planet in our solar system. It's so big that you could fit 1,300 Earths inside it! Jupiter is a gas giant, which means it's made mostly of hydrogen and helium, just like the Sun.

- **The Great Red Spot:** Jupiter has a giant red spot, which is actually a storm that has been raging for hundreds of years! This storm is so big that you could fit three Earths inside it!

- **Many Moons:** Jupiter has more than 75 moons! Some of them are even bigger than the planet Mercury.

- **Faint Rings:** Jupiter also has faint rings, but they're not as bright or as easy to see as Saturn's rings.

Fun fact: Jupiter's moon Ganymede is the largest moon in the solar system. It's even bigger than the planet Mercury!

Saturn: The Ringed Wonder

Saturn is the second-largest planet in our solar system, and it's famous for its beautiful rings. These rings are made of ice and rock, and they stretch out for thousands of miles.

- **The Rings:** Saturn's rings are made up of billions of tiny particles, ranging in size from grains of sand to huge boulders.

- **Many Moons:** Saturn also has many moons, including Titan, which has its own atmosphere and rivers and lakes of liquid methane.

- **Low Density:** Saturn is the least dense planet in our solar system. That means it's very light for its size. If there were a bathtub big enough, Saturn would float in it!

Fun fact: Saturn's moon Enceladus has geysers that shoot out plumes of water ice into space. Some scientists think there might be an ocean of liquid water beneath its icy surface.

Uranus: The Sideways Planet

Uranus is the seventh planet from the Sun, and it's tilted on its side, so it spins like a rolling ball! This unusual tilt gives Uranus some very strange seasons.

- **Tilted Axis:** Uranus's axis is tilted at almost 98 degrees, which means it's practically lying on its side.
- **Ice Giant:** Uranus is an ice giant, with a cold and windy atmosphere. It's made mostly of water, methane, and ammonia.

- **Rings and Moons:** Uranus also has rings and moons, but they're not as spectacular as Saturn's.

Fun fact: Uranus was the first planet discovered with a telescope. It was discovered by William Herschel in 1781.

Neptune: The Windy Planet

Neptune is the farthest planet from the Sun, and it's a dark and icy world. It has the strongest winds in the solar system, and it takes almost 165 Earth years to orbit the Sun once!

- **Strong Winds:** Neptune has winds that can reach speeds of over 1,200 miles per hour! That's faster than the speed of sound!

- **The Great Dark Spot:** Neptune used to have a Great Dark Spot, which was a giant storm similar to Jupiter's Great Red Spot. But this storm has since disappeared.

- **Triton:** Neptune has a large moon called Triton, which is unique because it orbits Neptune in the opposite direction of Neptune's rotation.

Fun fact: Neptune was the first planet to be discovered through mathematical prediction rather than observation.

Dwarf Planets - The Little Guys

Imagine a group of mini-planets, too small to be considered full-fledged planets but still fascinating in their own right. These are the dwarf planets, and they hang out in the outer regions of our solar system, beyond Neptune.

Pluto: The Most Famous Dwarf Planet

Pluto was once considered the ninth planet, but in 2006, it was reclassified as a dwarf planet. But don't feel sorry for Pluto – it's still a super interesting place!

- **Icy World:** Pluto is a very cold and icy world, located in a region of our solar system called the Kuiper Belt.

- **Five Moons:** Pluto has five moons! The largest, Charon, is so big compared to Pluto that they actually orbit each other like a double planet.

- **Heart-Shaped Feature:** Pluto has a large, heart-shaped feature on its surface called Tombaugh Regio. It's made of nitrogen ice and is thought to be a giant glacier.

Fun fact: Pluto is so far away from the Sun that it takes 248 Earth years to complete one orbit!

Other Dwarf Planets

Besides Pluto, there are four other officially recognized dwarf planets in our solar system:

- **Ceres:** Located in the asteroid belt between Mars and Jupiter, Ceres is the largest dwarf planet. It's so big that it makes up about one-third of the total mass of the asteroid belt!

- **Eris:** Eris is even farther away from the Sun than Pluto. It's a very cold and icy world, and it takes 557 Earth years to orbit the Sun once!

- **Makemake:** Makemake is also located in the Kuiper Belt, and it's covered in frozen methane. It's one of the brightest objects in the Kuiper Belt.

- **Haumea:** Haumea is a very unusual dwarf planet. It's shaped like a football, and it spins incredibly fast, completing one rotation in just under four hours!

Fun fact: Scientists think there may be hundreds or even thousands of other dwarf planets waiting to be discovered in the outer reaches of our solar system!

Planetary Atmospheres - Weather Across Worlds

Did you know that every planet in our solar system has a unique atmosphere? These layers of gases not only protect planets from space but also create incredible and sometimes extreme weather phenomena.

- **Mercury** has almost no atmosphere. Without a thick layer of gases, the surface experiences temperatures ranging from scorching hot during the day (800°F) to freezing cold at night (-290°F).

- **Venus** has a dense atmosphere made mostly of carbon dioxide. Its thick clouds trap heat, creating a "runaway greenhouse effect" that makes it the hottest planet in the solar system. Imagine sulfuric acid rain falling on a surface hot enough to melt lead!

- **Earth** is unique with its breathable mix of nitrogen and oxygen. Our atmosphere protects us from harmful space radiation and keeps our planet's temperature just right for life.

- **Mars** has a thin atmosphere of mostly carbon dioxide. This creates huge dust storms that can last for weeks, covering the entire planet!

- **Jupiter** has no solid surface, but its thick atmosphere of hydrogen and helium is home to storms larger than Earth. The Great Red Spot, for example, has been raging for over 300 years.

- **Saturn** is similar to Jupiter but has bands of strong winds that circle the planet, reaching up to 1,100 miles per hour! Its rings are made of ice and rock particles.

- **Uranus** has a pale blue-green color due to methane gas in its atmosphere. This icy planet is tipped on its side, leading to extreme and long seasons.

- **Neptune** is the windiest planet, with speeds reaching over 1,200 miles per hour. Its deep blue color also comes from methane, and its storms are some of the strongest in the solar system.

Fun Fact: Titan, Saturn's largest moon, has a thick atmosphere with rivers and lakes of liquid methane! Scientists think it's one of the best places to look for alien life.

Planetary Atmosphere Highlights

Planet	Atmosphere Composition	Weather Highlights
Mercury	Almost no atmosphere	Extreme temperatures
Venus	Carbon dioxide, sulfuric acid	Hottest planet, acid rain
Earth	Nitrogen, oxygen	Perfect for life, balanced climate
Mars	Carbon dioxide	Global dust storms
Jupiter	Hydrogen, helium	Great Red Spot, super storms
Saturn	Hydrogen, helium	Fast winds, icy rings

Planet	Atmosphere Composition	Weather Highlights
Uranus	Hydrogen, helium, methane	Tilted axis, extreme seasons
Neptune	Hydrogen, helium, methane	Fastest winds, deep blue storms

Asteroids and Comets - Space Rocks

Our solar system is also home to countless asteroids and comets, often called "space rocks." These objects are leftovers from the formation of the solar system, and they can tell us a lot about how our solar system came to be.

What are asteroids?

Asteroids are rocky objects that orbit the Sun. They come in all shapes and sizes, from small pebbles to huge boulders hundreds of miles wide. Most asteroids are located in the asteroid belt between Mars and Jupiter.

What are comets?

Comets are like giant snowballs made of ice and dust. They also orbit the Sun, but their orbits are often very long and elliptical. When a comet gets close to the Sun, the heat causes the ice to vaporize, creating a beautiful tail that can stretch for millions of miles.

Famous Asteroids and Comets

- **Halley's Comet:** This is probably the most famous comet. It's visible from Earth every 76 years, and it was last seen in 1986.

- **Ceres:** As we mentioned earlier, Ceres is now classified as a dwarf planet, but it was once considered the largest asteroid.

- **Bennu:** This asteroid is currently being studied by NASA's OSIRIS-REx spacecraft, which will bring a sample of Bennu back to Earth in 2023.

Fun facts about asteroids and comets:

- Some asteroids are large enough to have their own moons!

- Scientists think that an asteroid impact may have caused the extinction of the dinosaurs 66 million years ago.

- Comets are often called "dirty snowballs" because they're made of ice mixed with dust and rock.

Chapter 3:
Stars and Constellations
- A Night Sky Adventure

Have you ever looked up at the night sky and been amazed by the sheer number of stars?

They seem to twinkle and shimmer like tiny diamonds scattered across a dark velvet cloth.

But what exactly are stars, and why do they shine so brightly? Get ready to embark on a night sky adventure as we explore the fascinating world of stars and constellations

What Are Stars?

Stars are giant balls of hot gas, just like our Sun. They're incredibly hot and produce their own light and heat through a process called **nuclear fusion**. Imagine a giant furnace in space, burning brightly for billions of years!

- **Distance Makes Them Look Tiny**: Stars look tiny because they're incredibly far away from us. The nearest star to our solar system, besides the Sun, is **Proxima Centauri**, and it's still over **4 light-years** away! That means it takes light over 4 years to travel from that star to Earth.

- **Different Colors**: Stars come in different colors, from red to blue. The color of a star tells us about its temperature. **Red stars** are the coolest, while **blue stars** are the hottest. Our **Sun** is a yellow star, which means it's somewhere in the middle.

Different Types of Stars

Just like there are different types of fruit, there are different types of stars! Some are big, some are small, some are young, and some are old.

- **Giants and Supergiants**: Some stars are much bigger than our Sun. These are called **giants** and **supergiants**. Some can be over 1,000 times the size of the Sun! For example, **UY Scuti** is a supergiant star, about 1,700 times the size of our Sun.

- **Red Dwarfs**: These stars are smaller and cooler than our Sun. They burn their fuel very slowly and can live for up to trillions of years, far longer than most stars.

- **Binary Stars**: Some stars come in pairs, called **binary stars**. They orbit each other like dancers in space. For instance, **Alpha Centauri** is a binary system and the closest star system to Earth after the Sun.

- **White Dwarfs:** After medium-sized stars like our Sun burn out, they become **white dwarfs**. These stars are small and incredibly dense. Over billions of years, they slowly cool down.

The Life Cycle of a Star

Stars are born, they live for a long time, and then they die. It's like a story playing out in space!

- **Born in Nebulae:** Stars are born in giant clouds of gas and dust called **nebulae**. These clouds collapse under their own gravity, and the gas and dust clump together to form a **protostar**. The protostar gets hotter until it can start **nuclear fusion**.

- **Main Sequence Stars:** The protostar becomes a main-sequence star when it reaches the temperature needed to start nuclear fusion. This is the phase most stars, like our Sun, spend most of their lives in.

- **Red Giants:** As stars run out of fuel, they expand and cool down, becoming **red giants**. These stars are much larger than they were before and can be hundreds of times the size of their original self.

- **White Dwarfs:** After a red giant sheds its outer layers, it becomes a **white dwarf**, a dense star that slowly cools over billions of years.

- **Supernovae:** Massive stars end their lives in a **supernova**, a spectacular explosion that is so bright it can outshine an entire galaxy!

- **Neutron Stars and Black Holes:** After a supernova, the core can collapse to form a **neutron star** or a **black**

hole. Neutron stars are incredibly dense, while black holes have such strong gravity that even light can't escape them.

Constellations - Pictures in the Stars

For thousands of years, people have looked up at the stars and imagined patterns and pictures. These patterns are called **constellations**, and they've been used for navigation, storytelling, and even to track the seasons.

- **What Are Constellations?**: Constellations are groups of stars that seem to form shapes or patterns in the sky. They're like **connect-the-dots** pictures in space!

- **Famous Constellations**: Some of the most famous constellations include:

    o **Ursa Major** (the Big Dipper),

    o **Ursa Minor** (the Little Dipper),

    o **Orion** (the Hunter),

    o **Taurus** (the Bull).

- **Zodiac Constellations**: Have you ever heard of your **zodiac sign**? Like **Taurus the Bull** or **Leo the Lion**? Those are based on special constellations called the **zodiac constellations**. These 12 constellations lie along the path that the Sun seems to travel through the sky throughout the year.

Fun Fact: People have been fascinated by the zodiac for centuries, and they created stories and myths about each of these constellations.

- **Stories and Myths**: Many cultures have created stories and myths about the constellations. For example, the constellation **Orion** is named after a hunter in Greek mythology.

Finding Constellations

You can find constellations in the night sky by using **star charts** or **apps**. It's like a treasure hunt in space!

How to Use a Star Map:

- A star map shows the location of constellations visible at a particular time of year. You can use these charts to help guide your search for constellations.

Tips for Stargazing:

- The best time to observe the night sky is during the **new moon**, when the sky is darkest.

- **Light Pollution**: Avoid city lights for the best view of the stars. Head to a park or a remote area for the best stargazing experience.

Activity: Create a Constellation Book

Materials: Blank paper, markers, ruler, and stickers.
Instructions:

- Have kids pick their favorite constellations and create their own book or chart. They can draw each constellation and then write down fun facts or myths associated with it.

- As they learn about different constellations, they can add more pages to the book.

Activity: Make a Constellation Map

Materials: Cardboard, pushpins, a map of constellations, flashlight.

Instructions:

- Kids can create their own constellation map using pushpins on a piece of cardboard. They can look at a star chart and map out their favorite constellations.

- In a dark room, shine a flashlight through the pushpins to simulate the stars. This will help kids visualize the constellations in a fun and interactive way.

The Milky Way and Our Place in the Universe

The **Milky Way** is our galaxy, and it's made up of billions of stars, planets, and other objects. Our Solar System is just one small part of it.

Fun Fact: The Milky Way contains an estimated **100 billion stars**, and it's so large that light would take over **100,000 years** to travel from one side to the other.

Activity: Star Chart and Observation Log

Materials: Notebook, pen or pencil, star map, ruler.

Instructions:

- Keep a log of the constellations you spot. For each observation, draw the constellation and note the date and time.

- Over several months, you'll see how the constellations shift in the sky, which is a result of the Earth's orbit around the Sun.

Stars and Light Pollution

Light pollution from cities makes it hard to see stars and constellations. Find a dark place away from city lights for better stargazing.

Chapter 4:
Space Exploration - Boldly Going

Imagine blasting off from **Earth** in a powerful rocket, soaring through the atmosphere, and leaving our planet behind!

Space exploration is all about pushing the boundaries of human knowledge and daring to venture into the vast unknown. Let's discover how humans have reached for the stars and what exciting adventures lie ahead!

The History of Space Travel

Humans have always looked up at the stars and wondered what was out there. But it wasn't until the 20th century that we finally developed the technology to leave Earth and explore space.

- **Early Rockets:** The first rockets were invented in ancient China, but it took centuries of research and development before rockets were powerful enough to reach space. Pioneers like Robert Goddard and

Wernher von Braun made significant contributions to rocket technology.

- **The Space Race:** In the 1950s and 1960s, the United States and the Soviet Union engaged in a competition known as the Space Race. Both countries wanted to be the first to achieve major milestones in space, such as launching satellites and sending humans into orbit.

Firsts in Space:

- In 1957, the Soviet Union launched **Sputnik 1**, the first artificial satellite to orbit Earth. This marked the beginning of the Space Age.

- In 1961, Soviet cosmonaut **Yuri Gagarin** became the first human in space, orbiting Earth aboard the Vostok 1 spacecraft.

- In 1969, American astronauts **Neil Armstrong** and **Buzz Aldrin** became the first humans to walk on the Moon during the Apollo 11 mission.

Famous Astronauts and Their Missions

Many brave astronauts have ventured into space, each contributing to our understanding of the universe.

- **Valentina Tereshkova:** The first woman in space, she orbited Earth 48 times in 1963 aboard the Vostok 6 spacecraft.

- **Sally Ride:** The first American woman in space, she flew aboard the Space Shuttle Challenger in 1983.

- **Neil Armstrong:** As commander of Apollo 11, he took "one small step for man, one giant leap for mankind" when he became the first human to set foot on the

Moon.

- **Buzz Aldrin:** The second human to walk on the Moon, he piloted the lunar module during the Apollo 11 mission.

- **Chris Hadfield:** A Canadian astronaut known for his captivating videos and music performances from the International Space Station.

Important Space Missions

Throughout history, there have been countless space missions that have expanded our knowledge of the universe.

- **Apollo Missions:** The Apollo program was a series of missions that landed humans on the Moon between 1969 and 1972. These missions collected valuable scientific data and brought back lunar samples to Earth.

- **Voyager Probes:** Launched in 1977, the Voyager 1 and Voyager 2 probes are still exploring the outer reaches of our solar system. They have sent back stunning images of Jupiter, Saturn, Uranus, and Neptune, and they are now venturing into interstellar space.

- **Hubble Space Telescope:** Orbiting Earth since 1990, the Hubble Space Telescope has captured breathtaking images of distant galaxies, nebulae, and other celestial objects. It has revolutionized our understanding of the universe.

Timeline of Space Exploration

Humans have always dreamed of reaching for the stars. From launching the first satellite to planning missions to Mars, space exploration has come a long way. Here are some of the most exciting milestones:

- **1957: Sputnik 1** becomes the first artificial satellite to orbit Earth, launched by the Soviet Union. This marks the beginning of the Space Age.

- **1961:** Yuri Gagarin becomes the first human to travel into space, orbiting Earth aboard the Soviet spacecraft Vostok 1.

- **1969:** NASA's Apollo 11 mission lands the first humans on the Moon. Neil Armstrong's famous words—"That's one small step for man, one giant leap for mankind"—inspire the world.

- **1977:** The Voyager 1 and Voyager 2 spacecraft are launched to study the outer planets. They are now traveling in interstellar space, carrying golden records with information about Earth.

- **1990:** The Hubble Space Telescope is launched, providing stunning images of galaxies, nebulae, and more, revolutionizing our understanding of the universe.

- **2000:** The International Space Station (ISS) begins hosting astronauts, serving as a symbol of global collaboration in space.

- **2021:** NASA's Perseverance rover lands on Mars, searching for signs of ancient microbial life and collecting samples for future missions.

- **Future:** Plans are underway for human missions to Mars, lunar bases, and even space tourism. Imagine living and working on another planet!

Activity: Design Your Own Space Mission

Space missions take years of planning and teamwork. Now, it's your turn to create a mission to explore the cosmos!

What You'll Need:

- Paper and markers
- Imagination

Steps:

1. **Choose Your Destination:** Will your mission explore a nearby planet like Mars, a moon like Europa, or a distant galaxy?
2. **Name Your Mission:** Give your mission a cool and inspiring name.
3. **Design Your Spacecraft:** Draw a spacecraft that can complete the mission. Does it need solar panels, robotic arms, or tools to collect samples?
4. **Set Your Mission Goals:** Write down three main objectives for your mission. For example:
   - "Search for water on Europa."
   - "Study the atmosphere of Venus."
   - "Map the surface of an asteroid."
5. **Imagine Your Discoveries:** What exciting things might your mission find? New life forms? Rare minerals? Amazing photos of distant stars?

Fun Twist: Share your mission design with friends or family and vote on whose mission is the most adventurous!

The International Space Station

- The International Space Station (ISS) is a large spacecraft that orbits Earth. It's like a giant floating laboratory where astronauts from different countries live and work together, conducting scientific experiments and learning more about how to live in space.

- **International Collaboration:** The ISS is a joint project of many countries, including the United States, Russia, Europe, Japan, and Canada.

- **Life on the ISS:** Astronauts on the ISS experience microgravity, which means they float around! They have to eat special food, sleep in sleeping bags attached to the wall, and exercise regularly to stay healthy.

The Future of Space Exploration

The future of space exploration is full of exciting possibilities!

- **Missions to Mars:** Scientists are planning missions to send humans to Mars in the coming decades. Imagine being one of the first people to set foot on the Red Planet!

- **Space Tourism:** Companies like SpaceX and Blue Origin are developing spacecraft to take tourists into space. Maybe one day, you'll be able to take a vacation to orbit Earth or even visit the Moon!

- **Exploring Other Star Systems:** Scientists are also working on technologies that could allow us to travel to other star systems and explore planets beyond our solar system. Who knows what we might find out there?

Chapter 5:
Life in Space - Out of This World

Imagine floating like a feather, eating food that drifts in mid-air, and sleeping in a sleeping bag attached to the wall! Life in space is an extraordinary adventure, full of unique challenges and incredible experiences. Let's discover how astronauts live and work in this amazing environment.

How Do Astronauts Live in Space?

Living in space is very different from living on Earth. Astronauts have to adapt to a world without gravity, where the ordinary becomes extraordinary!

- **Microgravity:** In space, there's very little gravity, a condition called microgravity. This means astronauts float! Imagine lifting heavy objects with just a finger or doing somersaults in the air effortlessly. But microgravity also has its challenges. Astronauts need to learn how to move around, eat, and even sleep without things floating away.

- **Spacesuits:** Stepping outside a spacecraft requires a special outfit – a spacesuit! These high-tech suits are like personal spaceships, providing oxygen to breathe, regulating temperature, and protecting astronauts from harmful radiation and extreme temperatures. They even have built-in toilets!

- **Life Support Systems:** Spacecraft are like mini-Earths, equipped with life support systems that provide everything astronauts need to survive. These systems recycle air and water, control temperature, and even generate power from sunlight.

Daily Life in Orbit

- **Space Food:** Forget about crumbs and spills! Space food is specially designed to be eaten in microgravity. Think of bite-sized snacks, tortillas instead of bread, and lots of dehydrated foods that are rehydrated with water. Salt and pepper come in liquid form to prevent them from floating away and clogging air vents.

- **Sleeping in Space:** Astronauts sleep in special sleeping bags attached to the wall to prevent them from floating around and bumping into things. They also wear eye masks because the Sun rises and sets every 90 minutes in orbit!

- **Hygiene in Space:** Keeping clean in space can be tricky! Astronauts use special shampoos and soaps that don't need rinsing, and they use wet wipes for washing. Going to the bathroom in space is also a unique experience, involving special toilets with air flow to collect waste.

- **Staying Entertained:** Life in space isn't all work and no play. Astronauts have free time to read books, watch movies, play games, and even make video calls to their families back on Earth.

Challenges of Living in Space

While living in space is an incredible adventure, it also comes with its fair share of challenges.

- **Effects of Microgravity:** Microgravity can weaken muscles and bones because they don't have to work against gravity like they do on Earth. Astronauts have to exercise for at least two hours every day to stay strong and healthy. Special exercise machines help them work out in a weightless environment.

- **Psychological Challenges:** Spending long periods in a confined space, away from family and friends, can be mentally challenging. Astronauts need to be resilient, adaptable, and good at teamwork to cope with the isolation and stress of space travel.

- **Radiation Exposure:** Space is filled with harmful radiation that can increase the risk of health problems. Spacecraft and spacesuits provide some protection, but astronauts are still exposed to higher levels of radiation than they would be on Earth.

The Search for Extraterrestrial Life

One of the most captivating questions in space exploration is: Are we alone in the universe? Scientists are constantly searching for signs of life beyond Earth.

- **SETI:** The Search for Extraterrestrial Intelligence (SETI) uses powerful radio telescopes to listen for signals that might be sent by other civilizations in space. Imagine picking up a radio message from an alien world!

- **Exploring Extreme Environments:** Scientists are also studying extreme environments on Earth, like deep-sea vents and hot springs, to learn more about the types of life that might exist in the harsh conditions of other planets.

- **The Possibility of Life:** With billions of stars and planets in the universe, many scientists believe that life may exist elsewhere. Discovering even microbial life on another planet would be one of the greatest discoveries in human history!

Did you know? Astronauts often experience a "growth spurt" in space! Because of microgravity, their spines can lengthen, making them a little bit taller. But this effect is temporary, and they return to their normal height when they come back to Earth.

Chapter 6: Incredible Space Inventions

Space exploration is a thrilling adventure, a journey into the vast unknown. But it wouldn't be possible without the brilliant minds that invent and engineer the incredible tools and technologies that allow us to reach for the stars.

In this chapter, we'll delve into the fascinating world of space inventions, from powerful rockets to ingenious rovers, and discover how they help us unlock the secrets of the universe.

Rockets: Our Ride to Space

Imagine a machine so powerful that it can defy gravity, carrying humans and cargo beyond Earth's atmosphere and into the depths of space. That's a rocket! These incredible feats of engineering work by burning fuel to create hot gas, which is expelled out of a nozzle with tremendous force, propelling the rocket upward.

- **A Brief History:** The earliest rockets were invented in ancient China, using gunpowder as fuel. Over centuries, rockets evolved, becoming more sophisticated and powerful. Pioneers like Robert Goddard, a visionary American scientist, and Wernher von Braun, a German rocket engineer, made significant contributions to rocket technology, paving the way for modern space exploration.

- **Types of Rockets:** Rockets come in various shapes and sizes, each designed for specific missions.

    o **Sounding Rockets:** These smaller rockets are used to study Earth's upper atmosphere and conduct scientific experiments in microgravity.

    o **Orbital Rockets:** These powerful rockets launch satellites and spacecraft into orbit around Earth.

    o **Interplanetary Rockets:** These massive rockets, like the Saturn V that launched the Apollo missions, are designed to send spacecraft to other planets and moons in our solar system.

- **Stages to Success:** Many rockets are built in stages, like a layered cake. Each stage fires for a certain amount of time before being discarded to reduce weight and increase efficiency. This allows the rocket to reach incredible speeds and travel vast distances.

- **Fueling the Journey:** Rockets use different types of fuel, depending on their size and mission. Liquid propellants, like liquid hydrogen and liquid oxygen, are commonly used for larger rockets, while solid propellants, similar to those used in fireworks, are used for smaller rockets and boosters.

Satellites: Our Eyes in the Sky

Satellites are like artificial moons that orbit Earth, constantly observing and communicating. They are indispensable tools for a wide range of applications, from connecting people across the globe to unraveling the mysteries of the cosmos.

- **Communication Satellites:** These satellites are like giant relay stations in space, beaming signals for television, radio, and telephone communications around the world. They allow us to make phone calls, video chat with loved ones, watch live events, and access the internet from almost anywhere on Earth.

- **Navigation Satellites:** Ever used a GPS device or app to find your way? That's thanks to navigation satellites! These satellites constantly transmit signals that allow receivers on Earth to determine their precise location. They are used in everything from car navigation systems and aircraft guidance to tracking wildlife and monitoring natural disasters.

- **Weather Satellites:** These watchful eyes in the sky keep a constant vigil on Earth's weather patterns, providing valuable data to meteorologists who predict the weather. They can track storms, measure temperature and humidity, observe cloud formations, and even monitor changes in Earth's climate over time.

- **Scientific Satellites:** These specialized satellites are designed to conduct a wide range of scientific research. They study Earth's atmosphere, oceans, and landmasses, map the universe, search for exoplanets, and observe distant stars and galaxies.

Space Telescopes: Seeing the Invisible

Telescopes are like powerful magnifying glasses that allow us to see distant objects in space. But space telescopes, positioned high above Earth's atmosphere, have an even clearer view, free from the distortions caused by air and light pollution.

- **Hubble Space Telescope:** This iconic telescope, launched in 1990, has revolutionized our understanding of the universe. It has captured breathtaking images of distant galaxies, [1] nebulae, and other celestial objects, revealing the vastness and beauty of the cosmos.

- **James Webb Space Telescope:** This next-generation space telescope, set to launch soon, will be even more powerful than Hubble. With its advanced technology, it will be able to peer even farther into space, observing the first stars and galaxies that formed after the Big Bang and searching for signs of life on exoplanets.

Space Probes: Exploring Distant Worlds

Space probes are robotic explorers that venture into the depths of our solar system and beyond, acting as our eyes and ears in distant realms. These unmanned spacecraft carry scientific instruments to collect data about planets, moons, asteroids, and comets, sending valuable information back to Earth.

- **Voyager Probes:** Launched in 1977, the Voyager 1 and Voyager 2 probes are humanity's farthest-reaching emissaries. They have journeyed past Jupiter, Saturn, Uranus, and Neptune, sending back stunning images and groundbreaking data about these gas giants and their moons. Now, they are venturing into interstellar space, exploring the vast region between the stars.

- **Mars Rovers:** These intrepid rovers, like Curiosity and Perseverance, are traversing the rugged terrain of Mars, acting as robotic geologists and explorers. They are equipped with cameras, drills, and scientific instruments to analyze the Martian soil, rocks, and atmosphere, searching for signs of past or present life and paving the way for future human missions to the Red Planet.

Space Suits: Our Personal Spaceships

When astronauts venture outside their spacecraft, they need a special kind of protection – a spacesuit. These high-tech suits are like personal spaceships, providing everything an astronaut needs to survive in the harsh environment of space.

- **Life Support:** Space suits are equipped with sophisticated life support systems that provide oxygen to breathe, regulate temperature, and protect astronauts from harmful radiation and micrometeoroids, which are tiny, high-speed particles that can cause damage.

- **Mobility:** Imagine trying to move around in a bulky suit in the weightlessness of space! Space suits are designed to allow astronauts to move with flexibility and dexterity, with special joints and gloves that enable them to perform tasks like repairs and experiments.

- **Communication:** Clear communication is essential in space. Space suits have built-in communication systems that allow astronauts to talk to each other and to mission control on Earth, ensuring they can stay connected and receive support during their spacewalks.

Chapter 7: Space Phenomena and Mysteries

Space is full of wonders that can spark our curiosity and leave us in awe. In this chapter, we'll explore some of the most incredible and mysterious phenomena that occur in the vast expanse of the universe. Get ready to be amazed!

Black Holes - Cosmic Vacuum Cleaners

Imagine an object in space with such strong gravity that nothing, not even light, can escape its pull! That's a black hole, a region of space-time where gravity is so intense that it warps the fabric of the universe.

How Do Black Holes Form?

Black holes are often formed when massive stars collapse at the end of their lives. The star's core collapses in on itself, squeezing matter into an incredibly small space. This creates an extremely dense object with immense gravitational pull. When the core of a star is massive enough, it will eventually collapse under its own gravity, creating a singularity—a point where gravity is infinitely strong.

Event Horizon

The boundary around a black hole beyond which nothing can escape is called the event horizon. Once something crosses the event horizon, it's gone forever! This point marks the "point of no return." Anything that passes this threshold, including light, cannot escape the black hole's gravitational pull.

Types of Black Holes

There are different types of black holes:

- Stellar-mass Black Holes: These are formed when massive stars collapse, typically with a mass a few times that of our Sun.

- Supermassive Black Holes: These giants can be millions or even billions of times the mass of our Sun. They are found at the centers of galaxies, including our own Milky Way! These supermassive black holes are crucial in the formation and structure of galaxies.

Hawking Radiation

While black holes are known for trapping everything that enters them, in 1974, physicist Stephen Hawking proposed that black holes could actually emit radiation. This radiation, now known as Hawking Radiation, is the result of quantum effects near the event horizon. According to quantum mechanics, particle-antiparticle pairs can spontaneously form near the event horizon. One of these particles may fall into the black hole while the other escapes, leading to a slow loss of mass for the black hole. This process suggests that black holes might eventually evaporate over incredibly long timescales.

Seeing the Invisible

We can't see black holes directly because they don't emit light. However, we can detect them by observing their effects on nearby stars and gas. For example, when a black hole pulls in gas, it forms an accretion disk—a rotating disk of hot gas that emits X-rays. These X-rays can be detected by space telescopes, allowing scientists to infer the presence of a black hole.

The Fate of Black Holes

Because of their immense gravity, black holes continue to grow as they pull in matter from their surroundings. Over time, this growth leads to the formation of supermassive black holes at the centers of most galaxies. But with the concept of Hawking Radiation, black holes could theoretically "evaporate" over unimaginable timescales, slowly releasing energy until they eventually disappear.

Wormholes - Shortcuts Through Space-Time

Imagine a tunnel that connects two distant points in space-time, bending the very fabric of the universe. These hypothetical structures are known as **wormholes**. They are a key concept in Einstein's theory of **General Relativity** and have been popularized in science fiction.

What Are Wormholes?

Wormholes are sometimes described as "shortcuts" through space. If they exist, they could potentially allow us to travel between distant parts of the universe almost instantaneously, bypassing the vast distances between stars and galaxies.

Are Wormholes Real?

Currently, wormholes remain theoretical. While equations suggest they could exist, there is no empirical evidence to support their existence. Some theories propose that wormholes might be created by intense gravitational fields or may form as a result of quantum fluctuations. However, stabilizing a wormhole (to prevent it from collapsing) would require exotic matter with negative energy, something we have yet to discover.

Quasars - The Brightest Objects in the Universe

Quasars are incredibly bright, energetic, and distant objects powered by supermassive black holes at the centers of galaxies. They are among the most powerful and luminous objects known to humanity.

What Are Quasars?

A quasar is essentially a very energetic and active galactic nucleus, where a supermassive black hole is pulling in large amounts of gas and dust. As this material spirals into the black hole, it forms an **accretion disk**—a rotating disk of hot gas that emits immense amounts of radiation. This radiation is what makes quasars visible across vast distances.

The Power of Quasars

The energy released by quasars can outshine the entire galaxy that contains them, making them visible from billions of light-years away. The power of quasars is thought to be generated by the release of energy as matter is pulled into the black hole, heating up to millions of degrees and emitting electromagnetic radiation.

Space-Time and the Fabric of the Universe

The concept of **space-time** is central to our understanding of the universe. According to Albert Einstein's theory of **General Relativity**, space and time are not separate entities but are woven together into a single four-dimensional fabric. This fabric can be stretched, warped, and curved by massive objects.

General Relativity and Space-Time Curvature

Einstein's theory revolutionized our understanding of gravity. Instead of thinking of gravity as a force between masses, as Newton proposed, Einstein described it as the bending of space-time caused by the presence of mass and energy. Large objects like stars, planets, and black holes create indentations in the fabric of space-time, causing smaller objects to move toward them.

Gravitational Waves - Ripples in Space-Time

In 2015, scientists detected **gravitational waves** for the first time—ripples in space-time caused by the acceleration of massive objects, like the collision of two black holes. These waves were predicted by Einstein a century ago, and their detection opened up a new way of observing the universe. Gravitational waves allow us to study cosmic events that were previously invisible to traditional telescopes.

Time Dilation

One of the most mind-bending consequences of general relativity is time dilation. This phenomenon occurs when an object moves close to the speed of light or is near a massive gravitational field. In these cases, time for the object moves slower relative to an observer who is far away from the gravitational influence. Time dilation has been experimentally verified through measurements of atomic clocks on satellites and high-speed particles.

Supernovae - Stellar Explosions

When some stars reach the end of their lives, they go out with a bang! A supernova is a powerful explosion that marks the end of a star's life cycle.

- **Brilliant Light Show:** Supernovae are incredibly bright, sometimes outshining an entire galaxy for a short period. They release enormous amounts of energy and matter into space.

- **Creating Elements:** Supernovae are responsible for creating many of the elements that make up our bodies and the world around us, such as oxygen, iron, and gold. We are literally made of stardust!

- **Neutron Stars and Black Holes:** After a supernova, the core of the star can collapse to form a neutron star, an incredibly dense object that spins rapidly, or a black hole.

Dark Matter and Dark Energy - The Hidden Universe

When we look up at the night sky, it seems like we can see everything the universe has to offer. Stars, galaxies, planets, and even distant nebulae shine brightly against the darkness. But did you know that most of the universe is invisible to us?

Scientists estimate that only about **5% of the universe** is made up of ordinary matter—the stuff we can see and touch. The rest is a mystery! About **27%** is something called dark matter, and a whopping **68%** is dark energy. These two mysterious forces play a huge role in shaping the universe, even though we can't see them directly.

What Is Dark Matter?

Dark matter is like an invisible glue that holds galaxies together. Without it, galaxies would fly apart because there isn't enough visible matter to create the gravity needed to keep them intact.

- **How Do We Know It's There?**
  Scientists can't see dark matter directly, but they observe its effects. For example, when they look at the way galaxies rotate, they notice that stars on the edges are moving faster than they should. This can only be explained if there's extra invisible mass—dark matter—creating more gravity.

- **Where Is Dark Matter?**
  Dark matter is everywhere in the universe, forming a web-like structure that connects galaxies. If we could see it, space would look like a glowing spiderweb of stars and galaxies!

What Is Dark Energy?

Dark energy is even more mysterious than dark matter. It's a force that makes the universe expand faster and faster over time, like a balloon being inflated.

- **How Do We Know It Exists?**
  In 1998, scientists discovered that distant galaxies are moving away from us at an accelerating rate. This means some kind of energy—dark energy—is pushing them apart.
- **Why Is It Important?**
  Dark energy controls the fate of the universe. Will it expand forever? Will it slow down and stop? Or will it collapse back in on itself? Scientists are still trying to figure out the answers.

Fun Fact: If you could see dark matter and dark energy, the night sky would look completely different! Galaxies would glow against a shimmering backdrop of mysterious energy.

Other Fascinating Phenomena

1. **Gamma-Ray Bursts:**
   The universe's most powerful explosions, caused by the collapse of massive stars or the collision of neutron stars. In just seconds, they release more energy than our Sun will in its entire lifetime!

2. **Gravitational Waves:**
   Ripples in the fabric of space and time, predicted by Einstein and first detected in 2015. These waves are created by massive events, like black holes colliding.

3. **The Multiverse Theory:**
   Could there be more than one universe? Some scientists think so! The multiverse theory suggests that our universe might be just one of countless others, each with its own unique laws of physics.

Activity: Map the Invisible Universe

What You'll Need:

- Black paper
- White chalk or a marker
- Imagination

Steps:

1. Draw the galaxies you can "see" on the black paper. These might look like small spirals, dots, or clusters.
2. Now imagine the dark matter that connects these galaxies. Use your chalk or marker to draw web-like lines linking the galaxies together.
3. Add a glowing aura around your drawing to represent dark energy expanding the universe.
4. Share your map with friends or family and explain how the invisible universe works!

Nebulae - Stellar Nurseries

Nebulae are vast clouds of gas and dust in space. They are often called "stellar nurseries" because they are the birthplaces of stars.

- **Colorful Clouds:** Nebulae come in a variety of stunning colors and shapes. The different colors are caused by different elements in the gas, such as hydrogen, helium, and oxygen.

- **Famous Nebulae:** Some famous nebulae include the Orion Nebula, a stellar nursery visible to the naked eye, the Eagle Nebula, with its iconic "Pillars of Creation," and the Carina Nebula, home to some of the most massive stars in our galaxy.

- **Planetary Nebulae:** When a star like our Sun dies, it can create a beautiful expanding shell of gas called a planetary nebula. These glowing clouds of gas disperse into space, enriching the interstellar medium with the building blocks for new stars.

Auroras - Dancing Lights in the Sky

Have you ever seen the northern lights or southern lights? These colorful displays of light in the night sky are called auroras.

- **Solar Wind:** Auroras are caused by charged particles from the Sun, called the solar wind, interacting with Earth's magnetic field and atmosphere. These particles excite the atoms in the atmosphere, causing them to emit light.

- **Dancing Lights:** The auroras appear as shimmering curtains, ribbons, or rays of light that dance across the sky in shades of green, red, blue, and purple. They are a breathtaking sight!

- **Where to See Them:** Auroras are most often seen near the poles, in regions like Alaska, Canada, and Scandinavia (northern lights) or Antarctica and southern parts of Australia and New Zealand (southern lights).

Cosmic Collisions!

- **Asteroid Impacts:** Remember those "space rocks" we talked about? Sometimes, asteroids collide with planets and moons, creating massive craters. Scientists believe a huge asteroid impact wiped out the dinosaurs millions of years ago!

- **Galaxy Mergers:** Even galaxies can collide! When galaxies collide, it's a slow and spectacular dance that can last for billions of years. Stars are flung around, gas clouds collide, and new stars are born. Our own Milky Way galaxy is on a collision course with the Andromeda galaxy, but don't worry, it won't happen for another few billion years!

Other Strange and Wonderful Space Phenomena

The universe is full of other fascinating and mysterious phenomena, such as:

- **Gamma-ray bursts:** The most powerful explosions in the universe, thought to be caused by the collapse of massive stars or the collision of neutron stars. These bursts release incredible amounts of energy in a short amount of time.

- **Gravitational waves:** Ripples in the fabric of spacetime caused by the acceleration of massive objects, like black holes or neutron stars. These waves were predicted by Einstein and were first detected in 2015, opening a new window into the universe.

- **Dark matter and dark energy:** Mysterious substances that make up most of the universe but cannot be seen directly. Scientists are still trying to understand what they are and how they work.

Chapter 8: Space Careers - Your Journey to the Stars

Have you ever wondered what it's like to work in space? From astronauts to engineers and even scientists searching for life on other planets, there are many exciting careers in space exploration.

Whether you dream of flying a spacecraft, studying the stars, or even designing rockets, the possibilities are endless. Let's take a look at some of the most amazing space jobs that could take you beyond Earth!

1. Astronaut - The Ultimate Explorer

Astronauts are the brave men and women who travel to space to explore, conduct experiments, and learn about the universe. They live and work on the International Space Station (ISS), orbiting Earth, or embark on missions to the Moon, Mars, and beyond.

- **What Do They Do?**
  Astronauts perform experiments, repair satellites, and even conduct spacewalks to maintain spacecraft. They

also test new technology to help future space explorers.

- **How to Become One:**
Becoming an astronaut requires years of education and training. Most astronauts are scientists, engineers, or medical doctors with strong problem-solving skills. They undergo intense physical training to survive the challenges of space.

Fun Fact: The first woman in space, Valentina Tereshkova, orbited Earth 48 times in 1963!

2. Aerospace Engineer - Designing the Future of Space Travel

Aerospace engineers design rockets, spacecraft, and everything needed to explore space. They work on projects like sending satellites into orbit, building spacecraft for astronauts, and even developing plans for interplanetary travel.

- **What Do They Do?**
Aerospace engineers use advanced technology to design the parts of rockets and spacecraft that keep astronauts safe and carry out missions. They also work on robotics, satellite systems, and space telescopes.

- **How to Become One:**
Aerospace engineers usually need a degree in aerospace engineering or mechanical engineering, followed by internships with space agencies or private companies like NASA or SpaceX.

Fun Fact: The Saturn V rocket, which took astronauts to the Moon, was built by thousands of aerospace engineers!

3. Astrobiologist - Searching for Life Beyond Earth

Astrobiologists study the possibility of life on other planets. They research environments where life could exist, such as the icy moons of Jupiter or the surface of Mars.

- **What Do They Do?**
  Astrobiologists study the conditions that support life on Earth and try to find similar conditions on other planets. They also examine meteorites and space dust for signs of life.

- **How to Become One:**
  To become an astrobiologist, you'll need a strong background in biology, chemistry, and physics. Many astrobiologists have PhDs in biology or astronomy and work with space agencies to conduct experiments.

  Fun Fact: One of the greatest discoveries astrobiologists hope for is finding microbial life on Mars!

4. Space Pilot - Flying High in Outer Space

Space pilots are highly trained individuals who pilot spacecraft that transport astronauts to space and bring them back safely. They operate the systems on spacecraft, ensuring they safely launch, navigate, and land.

- **What Do They Do?**
  Space pilots control the spacecraft, monitor its systems, and help astronauts during their space mission. They also work on training missions and help develop new spaceflight technology.

- **How to Become One:**
Becoming a space pilot requires training as a pilot and a background in aerospace engineering or physics. You also need to be highly skilled at handling aircraft and spacecraft in different conditions.

Fun Fact: The first space shuttle pilot was Robert Crippen, who flew the space shuttle Columbia in 1981!

5. Space Scientist - Unlocking the Secrets of the Universe

Space scientists explore the mysteries of space, studying planets, stars, black holes, and everything in between. They use telescopes, space probes, and data from space missions to learn more about the universe.

- **What Do They Do?**
Space scientists collect and analyze data to understand how planets form, how stars die, and how the universe began. They also study the effects of space on human health and how we can travel safely to other planets.

- **How to Become One:**
Becoming a space scientist usually requires a degree in astronomy, physics, or engineering. Many scientists pursue graduate studies and work at observatories or research labs.

Fun Fact: The Hubble Space Telescope has helped scientists discover thousands of new galaxies and planets!

6. Space Tourism - The New Frontier

Imagine one day taking a vacation to space! Space tourism is an emerging industry that allows civilians to travel to space as tourists. Companies like SpaceX,

Blue Origin, and Virgin Galactic are already working on making space tourism a reality.

- **What Do They Do?**
Space tourism companies design spacecraft and experiences for civilians, ranging from suborbital flights to possible moon missions. They also train tourists in space travel, including how to adjust to zero gravity.

- **How to Become One:**
Space tourists need to pass medical exams to ensure they're fit for space travel. Some companies are offering suborbital flights for those who can afford them, but more affordable options may become available in the future.

Fun Fact: The first civilian spaceflight took place in 2001 when Dennis Tito, an American businessman, flew to the ISS!

Activity: Design Your Space Career

What You'll Need:

- Paper and markers
- Creativity!

Steps:

1. Choose your dream space career: astronaut, engineer, scientist, or even space tourist!
2. Draw what your workspace or spacecraft might look like. Do you want to be working on Mars, exploring the Moon, or in a space laboratory?
3. Write down what tools or technology you would need to succeed in your career.
4. Share your design with your friends or family—maybe they'll be inspired to design their own space career!

Chapter 9:
Zodiac Signs - Unraveling the Mysteries of the Starry Sky

Have you ever wondered why people say things like, "Oh, he's such a Taurus!" or "She's a typical Libra"? These statements are connected to the ancient practice of astrology and the Zodiac! Let's dive into this fascinating world and explore the secrets behind these celestial signs.

What are Zodiac Signs?

For thousands of years, people have looked up at the stars and noticed patterns. Some of these patterns formed shapes that reminded them of animals, objects, or mythical heroes. These star patterns are called **constellations**.

Imagine a special path in the sky that the Sun seems to follow throughout the year. Along this path, we find 12 special constellations, and these are the Zodiac constellations! The word "Zodiac" comes from the Greek word *zōdiakos*, which means "circle of animals." Many of these constellations are named after animals, like Leo the Lion, Taurus the Bull, and Pisces the Fishes.

In astrology, each of these constellations is associated with a specific time of year. Astrologers believe that the position of the Sun, Moon, and planets among these constellations at the time of your birth can influence your personality and even your destiny.

The 12 Zodiac Constellations: Personality and Traits

Let's meet the 12 members of the Zodiac club and discover what makes each one unique!

Aries (The Ram): March 21 - April 19. Aries is the first sign of the Zodiac, and people born under this sign are known for being energetic, adventurous, and courageous, just like a ram charging fearlessly into the unknown! They are natural leaders, always ready to take on new challenges. Aries are also known for their enthusiasm, optimism, and competitive spirit.

Taurus (The Bull): April 20 - May 20. Taurus is an earth sign, and people born under this sign are known for being grounded, reliable, and patient. Like a bull, they are strong and determined, but they also have a gentle and loving side. Taureans appreciate the finer things in life, like good food, music, and art.

Gemini (The Twins): May 21 - June 20. Gemini is an air sign, and people born under this sign are known for being curious, communicative, and adaptable. Like twins, they have many different sides to their personality, and they love to learn new things and explore new ideas. Geminis are also known for their wit, charm, and social butterfly nature.

Cancer (The Crab): June 21 - July 22. Cancer is a water sign, and people born under this sign are known for being nurturing, sensitive, and compassionate. Like a crab with its hard shell, they can be protective of themselves and those they love. Cancers are also known for their strong intuition, creativity, and love of home and family.

Leo (The Lion): July 23 - August 22. Leo is a fire sign, and people born under this sign are known for being confident, creative, and generous. Just like a lion, they are natural leaders, and they love to be the center of attention. Leos are also known for their warmth, enthusiasm, and dramatic flair.

Virgo (The Virgin): August 23 - September 22. Virgo is an earth sign, and people born under this sign are known for being analytical, organized, and practical. They have a keen eye for detail and a love of order and efficiency. Virgos are also known for their intelligence, kindness, and dedication to helping others.

Libra (The Scales): September 23 - October 22. Libra is an air sign, and people born under this sign are known for being diplomatic, fair-minded, and harmonious. Like the scales of justice, they seek balance and fairness in all things. Libras are also known for their charm, grace, and love of beauty and art.

Scorpio (The Scorpion): October 23 - November 21. Scorpio is a water sign, and people born under this sign are known for being passionate, resourceful, and determined. Like a scorpion, they can be a bit mysterious and intense, but they are also fiercely loyal and protective of those they love. Scorpios are known for their strong will, intuition, and ability to see beneath the surface.

Sagittarius (The Archer): November 22 - December 21. Sagittarius is a fire sign, and people born under this sign are known for being optimistic, adventurous, and freedom-loving. Like an archer aiming for the stars, they are always looking ahead to new horizons and seeking knowledge and truth. Sagittarians are also known for their humor, honesty, and philosophical nature.

Capricorn (The Goat): December 22 - January 19. Capricorn is an earth sign, and people born under this sign are known for being responsible, disciplined, and ambitious. Like a mountain goat, they are determined to reach the top and achieve their goals. Capricorns are also known for their patience, practicality, and loyalty.

Aquarius (The Water Bearer): January 20 - February 18. Aquarius is an air sign, and people born under this sign are known for being independent, innovative, and humanitarian. They like to do things their own way and are often ahead of their time. Aquarians are also known for their intelligence, originality, and concern for the greater good.

Pisces (The Fishes): February 19 - March 20. Pisces is a water sign, and people born under this sign are known for being imaginative, compassionate, and artistic. Like fish swimming in the sea, they are often in their own dream world, full of creativity and empathy. Pisceans are also known for their sensitivity, intuition, and spiritual nature.

Zodiac Signs and Their Dates

As the Earth travels around the Sun, the Sun appears to pass through each of the Zodiac constellations. The time of year when the Sun is "in" a particular constellation determines the Zodiac sign for people born during that time.

Myths and Legends: The Stories Behind the Stars

The Zodiac constellations are not just patterns in the sky; they are also characters in fascinating myths and legends, often from ancient Greek and Roman mythology. These stories can tell us more about the symbolism and characteristics associated with each sign.

- **Aries the Ram:** In Greek mythology, Aries was a ram with a golden fleece. This ram was sent by the gods to save two children, Phrixus and Helle, from being sacrificed. The golden fleece later became the object of a quest by the hero Jason and the Argonauts. This myth symbolizes Aries' courage, determination, and willingness to take risks.

- **Taurus the Bull:** Taurus is associated with the story of Zeus, the king of the gods, who transformed himself into a bull to kidnap the beautiful princess Europa. He carried her across the sea to the island of Crete, which is how Europe got its name. This myth highlights Taurus' strength, passion, and appreciation for beauty.

- **Gemini the Twins:** Gemini represents the twins Castor and Pollux in Greek mythology. Castor was mortal, while Pollux was immortal. When Castor died, Pollux was so heartbroken that he begged Zeus to let him share his immortality with his brother. Zeus agreed, and they were placed together in the sky as the constellation Gemini. This myth symbolizes the duality of Gemini's nature, their close bonds with others, and their adaptability.

- **Cancer the Crab:** In the myth of Heracles (Hercules), Cancer was a giant crab sent by the goddess Hera to hinder Heracles during his battle with the Hydra. Although the crab was crushed under Heracles' foot, Hera placed it in the sky as a reward for its loyalty. This myth reflects Cancer's protective nature, tenacity, and connection to home and family.

- **Leo the Lion:** Leo is associated with the Nemean Lion, a ferocious beast with impenetrable skin that terrorized the countryside. Heracles, as part of his twelve labors, slew the lion and wore its skin as a trophy. This myth symbolizes Leo's strength, courage, and leadership qualities.

- **Virgo the Virgin:** Virgo is often associated with the goddess of agriculture and harvest, Demeter. She represents purity, service, and the abundance of nature. This myth highlights Virgo's practicality, attention to detail, and dedication to helping others.

- **Libra the Scales:** Libra is associated with Themis, the goddess of justice and balance. She represents fairness, harmony, and the pursuit of equality. This myth reflects Libra's diplomatic nature, their ability to see both sides of every story, and their desire for peace and harmony.

- **Scorpio the Scorpion:** In Greek mythology, Scorpio was sent by the goddess Artemis to kill the hunter Orion after he boasted that he would hunt every animal on Earth. The scorpion stung Orion, causing his death. This myth symbolizes Scorpio's intensity, passion, and ability to transform.

- **Sagittarius the Archer:** Sagittarius is often depicted as a centaur, a mythical creature with the upper body of a human and the lower body of a horse. Sagittarius is often associated with Chiron, the wise centaur who mentored many heroes. This myth highlights Sagittarius' love of freedom, their pursuit of knowledge, and their adventurous spirit.

- **Capricorn the Goat:** Capricorn is often depicted as a sea-goat, a creature with the head and upper body of a goat and the tail of a fish. In Greek mythology, Capricorn is associated with Pan, the god of the wild, who transformed himself into a sea-goat to escape the monster Typhon. This myth symbolizes Capricorn's ambition, resourcefulness, and ability to overcome challenges.

- **Aquarius the Water Bearer:** Aquarius is often associated with Ganymede, a beautiful young man who was kidnapped by Zeus to be the cupbearer of the gods. This myth highlights Aquarius' humanitarian nature, their independent spirit, and their desire to make the world a better place.

- **Pisces the Fishes:** In Greek mythology, Aphrodite and her son Eros transformed themselves into fish to escape the monster Typhon. They tied themselves together with a cord so they wouldn't lose each other in the vast ocean. This myth symbolizes Pisces' compassion, empathy, and connection to the spiritual realm

Chapter 9:
Beyond Our Solar System

We've explored our solar system, our Sun, planets, and moons. We've learned about stars and constellations. But the universe is far vaster than just our little corner of it! In this chapter, we'll journey beyond our solar system to discover the wonders that lie in the depths of space.

Exoplanets: Worlds Orbiting Other Stars

Did you know that our Sun isn't the only star with planets? In fact, there are billions of stars in our galaxy, the Milky Way, and many of them have their own planets orbiting them. These planets are called exoplanets, or extrasolar planets.

- **Discovering Distant Worlds:** Astronomers have developed incredible techniques to find exoplanets, even though they are very far away and difficult to see directly. One way is to look for tiny wobbles in a star's motion caused by the gravitational pull of an orbiting planet. Another way is to look for dips in a star's brightness when a planet passes in front of it, blocking some of its light.

- **A Variety of Worlds:** Exoplanets come in all shapes and sizes. Some are gas giants like Jupiter, while others are rocky like Earth. Some are scorching hot, while others are icy cold. Some orbit their stars very closely, while others are much farther away.

- **The Search for Life:** One of the most exciting things about exoplanets is the possibility that some of them might harbor life. Scientists are searching for exoplanets that have conditions similar to Earth, such as liquid water and a suitable atmosphere, where life as we know it could potentially exist.

Galaxies: Islands of Stars

Our solar system is just one tiny part of a much larger structure called a galaxy. A galaxy is a vast collection of stars, gas, dust, and planets, all held together by gravity.

- **The Milky Way:** Our galaxy is called the Milky Way, and it's a spiral galaxy, shaped like a giant pinwheel. It contains hundreds of billions of stars, including our Sun. If you look up at the night sky in a dark place, you can see a faint band of light stretching across the sky. That's the Milky Way!

- **Types of Galaxies:** Galaxies come in different shapes and sizes. Besides spiral galaxies like the Milky Way, there are elliptical galaxies, which are oval-shaped, and irregular galaxies, which have no defined shape.

- **Vast Distances:** The distances between galaxies are mind-bogglingly huge. The nearest major galaxy to our own is the Andromeda galaxy, and it's about 2.5 million light-years away! That means it takes light 2.5 million years to travel from Andromeda to Earth.

Exploring Our Galactic Neighbors: The Andromeda Galaxy

The Andromeda Galaxy, also known as M31, is the closest spiral galaxy to our Milky Way, located approximately 2.5 million light-years away. It's an immense collection of stars, gas, and dust—more than one trillion stars to be exact. Andromeda is easily visible to the naked eye from Earth, appearing as a faint smudge of light in the night sky, making it a perfect target for amateur astronomers.

The Collision of the Giants: Milky Way vs. Andromeda

One of the most exciting aspects of Andromeda is its predicted collision with the Milky Way in about 4.5 billion years. The two galaxies are on a collision course, moving toward each other at a speed of around 250,000 miles per hour (400,000 kilometers per hour). However, this "collision" won't look like a violent crash. Since stars are so far apart, it's unlikely that any stars will actually collide. Instead, the two galaxies will pass through each other, eventually merging into a single, larger galaxy.

This event will drastically alter the structure of both galaxies, resulting in a new galactic shape. Scientists predict that the new galaxy may form a giant elliptical galaxy, and our Sun could be moved to a new position. Though this collision won't happen for billions of years, it has sparked interest in the study of galactic mergers and their impact on star formation.

Other Fascinating Galaxies

While Andromeda is our closest neighbor, there are many other spectacular galaxies in the universe. Here are a few notable ones:

1. The Whirlpool Galaxy (M51):
   Located about 23 million light-years away, the Whirlpool Galaxy is a classic example of a spiral galaxy. It is currently in the process of merging with a smaller galaxy, creating a striking visual of two galaxies intertwined. The Whirlpool Galaxy is known for its

brilliant spiral arms and massive star-forming regions, making it one of the most photographed galaxies by astronomers.

2. The Sombrero Galaxy (M104):
    About 28 million light-years away, the Sombrero Galaxy gets its name from its resemblance to a wide-brimmed hat. It is a spiral galaxy with a large central bulge and a prominent disk surrounded by a halo of stars. The Sombrero is particularly famous for its well-defined dust lane, which can be clearly seen in photographs.

3. The Triangulum Galaxy (M33):
    At around 3 million light-years away, the Triangulum Galaxy is the third-largest member of our Local Group of galaxies (which includes the Milky Way and Andromeda). It's a smaller, less structured spiral galaxy with vast star-forming regions, similar to our own Milky Way.

4. The Large and Small Magellanic Clouds:
    These are two irregular galaxies that are part of the Milky Way's galactic neighborhood. The Large Magellanic Cloud is roughly 160,000 light-years away, while the Small Magellanic Cloud is around 200,000 light-years away. Both galaxies are irregular in shape and have been interacting with the Milky Way for billions of years, causing them to appear distorted.

Galactic Collisions: A Common Occurrence

Galactic collisions are not as rare as we might think. Over billions of years, galaxies are constantly interacting with one another, either through close passes or direct collisions. These events are crucial to the evolution of galaxies, often triggering bursts of star formation.

- How Collisions Affect Galaxies:

  When galaxies collide, their gas clouds can compress, causing new stars to form at an accelerated rate. However, the actual stars themselves rarely collide because galaxies are made up of mostly empty space. Instead, the gravitational forces of the collision can reshape the galaxies, warp their structures, and even lead to the creation of new types of galaxies.

- The Future of Our Milky Way:

  In addition to the Andromeda collision, our Milky Way is also on a collision course with the smaller Sagittarius Dwarf Galaxy and the Large Magellanic Cloud. These interactions, though less dramatic than the Andromeda collision, will still alter the structure of our galaxy over time.

The Fate of the Universe: Expanding and Collapsing Galaxies

The study of galaxies doesn't just focus on their collisions. Understanding how galaxies evolve is key to understanding the fate of the universe itself. The expansion of the universe—first discovered by Edwin Hubble in the 1920s—means that galaxies are gradually moving away from each other. This expansion is accelerating due to mysterious forces like dark energy.

However, as galaxies move farther apart, the universe could eventually face one of two fates:

1. The Big Freeze: The universe will continue to expand, with galaxies drifting farther apart, eventually reaching a state where stars burn out, and the universe becomes cold and dark.

2. **The Big Crunch:** Alternatively, the gravitational pull of matter in the universe could reverse the expansion, causing galaxies to collapse back together in a cataclysmic event.

We may not know which scenario will play out, but studying galaxies, their collisions, and their evolution provides us with a deeper understanding of the universe's past and future.

The Expanding Universe: A Journey Through Time

The universe is not static; it's constantly expanding! This means that galaxies are moving away from each other, like dots on a balloon that is being inflated.

- **The Big Bang:** Scientists believe that the universe began with a massive explosion called the Big Bang about 13.8 billion years ago. Since then, the universe has been expanding and cooling.

- **Looking Back in Time:** When we look at distant galaxies, we are seeing them as they were billions of years ago, because the light from those galaxies has taken billions of years to reach us. It's like looking back in time!

- **The Fate of the Universe:** Scientists are still trying to understand the ultimate fate of the universe. Will it continue to expand forever, or will it eventually collapse back on itself in a "Big Crunch"?

The Search for Extraterrestrial Life

One of the most captivating questions that has fascinated humanity for centuries is: **Are we alone in the universe?** With billions of stars and planets scattered across the cosmos, many scientists believe that the odds are in favor of discovering life beyond Earth. The search for extraterrestrial life, whether microbial or intelligent, has become one of the most exciting endeavors in modern science.

SETI - The Search for Extraterrestrial Intelligence

The **Search for Extraterrestrial Intelligence (SETI)** is an ongoing scientific effort to detect signals from alien civilizations. SETI researchers use radio telescopes to listen for unusual signals, hoping to detect patterns that could indicate intelligent life. The most famous SETI project, the **Arecibo Message**, was sent into space in 1974, aiming to reach distant civilizations with a binary-coded message. While SETI has yet to discover any signals from aliens, the search continues, and every new discovery brings us closer to answering this age-old question.

How Does SETI Work?

SETI operates by scanning the sky for radio waves or signals that seem unusual or purposeful. Scientists compare these signals to natural cosmic background noise, looking for signs of intelligence. However, detecting alien signals is incredibly challenging due to the vast distances between stars and the possibility that extraterrestrial civilizations may use different forms of communication.

UFOs - Unidentified Aerial Phenomena

In addition to scientific searches, reports of **Unidentified Flying Objects (UFOs)**—now referred to as **Unidentified Aerial Phenomena (UAPs)**—have been a topic of intrigue for decades. Governments, researchers, and amateur enthusiasts have all reported sightings of mysterious objects in the sky.

Government and Military Encounters

In recent years, there have been declassified reports from the U.S. government and military acknowledging the existence of UAPs. These reports have reignited the debate about whether these phenomena could be evidence of extraterrestrial technology or unknown natural phenomena. While no direct evidence of alien spacecraft has been found, many believe that these reports may be a step toward uncovering the truth.

The Fermi Paradox - Where Are They?

The **Fermi Paradox** highlights the contradiction between the high probability of extraterrestrial civilizations existing and the lack of evidence or contact with such civilizations. Given the billions of potentially habitable planets in the Milky Way alone, scientists are left to wonder why we haven't detected any signs of life.

Possible Solutions to the Paradox

- **The Great Filter:** One theory suggests that there may be a "Great Filter" in the development of intelligent life. This filter could be a stage in evolution that is so difficult to pass that most civilizations never make it to the point where they can communicate or travel through space.
- **They're Out There, but We're Not Listening Right:** Another theory is that alien civilizations are indeed out there, but they might be using technologies or methods of communication that we don't understand or are not yet able to detect.
- **The Zoo Hypothesis:** Some suggest that extraterrestrial civilizations are aware of us but intentionally avoid contact, much like humans observing animals in a zoo. They may be waiting for us to evolve further or reach a certain technological stage before they make contact.

Extraterrestrial Life on Other Worlds

In addition to the search for intelligent life, scientists are also focusing on finding microbial life in places like **Mars**, the **Moons of Jupiter and Saturn**, and beyond.

Mars - The Red Planet's Potential for Life

Mars, our neighboring planet, has long been a target in the search for life. Evidence of ancient riverbeds, lakes, and minerals that form in the presence of water suggests that Mars may have supported life in the past. NASA's **Perseverance Rover** is currently exploring the planet's surface to collect samples and search for signs of ancient microbial life.

The Moons of Jupiter and Saturn - Europa and Enceladus

The icy moons of **Jupiter** (Europa) and **Saturn** (Enceladus) are two of the best places to search for extraterrestrial life. Both moons are believed to have subsurface oceans beneath their icy crusts, which could harbor microbial life. Scientists are eager to send missions to these moons to explore their potential for life.

The Drake Equation - Estimating the Odds

In 1961, astronomer **Frank Drake** developed the **Drake Equation**, a formula designed to estimate the number of technologically advanced civilizations in our galaxy. The equation considers factors like the rate of star formation, the number of planets that could support life, and the likelihood of life developing on those planets. While the exact numbers are still uncertain, the equation suggests that there could be hundreds or even thousands of civilizations in the Milky Way alone.

Are We Ready for Contact?

The discovery of extraterrestrial life would have profound implications for humanity. It would challenge our understanding of our place in the universe and raise questions about how we would interact with other civilizations. While the discovery of alien life has not yet occurred, the search continues, and with new technologies and missions, we may be closer than ever to finding the answer to the question: **Are we alone in the universe?**

Space Activities

1. Constellation Viewer

- **Materials:** Empty cardboard tube (like from a paper towel roll), scissors, black marker, pushpin, flashlight
- **Instructions:**

    1. Help your child find a simple constellation map online or in a book. Choose a constellation that is visible in the night sky at your current location and time of year.

    2. Use the black marker to draw the constellation pattern on one end of the cardboard tube. Make the dots big enough to be easily visible.

    3. Carefully use the pushpin to poke holes where the stars are on the drawing. The bigger the hole, the brighter the "star" will appear.

    4. In a dark room, shine the flashlight through the open end of the tube onto a wall or ceiling. The light shining through the holes will create a mini constellation projection!

    5. Experiment with different constellations and see if you can identify them in the real night sky.

2. Solar System Mobile

- **Materials:** Hangers, string or yarn, different sized Styrofoam balls, paint, markers, glitter (optional), small pictures of the planets

- **Instructions:**

  1. Gather different sized Styrofoam balls to represent the planets. Use the smallest for Mercury and the largest for Jupiter.

  2. Paint each Styrofoam ball to represent the different planets in our solar system. Use your imagination and add details like rings for Saturn or the Great Red Spot for Jupiter.

  3. Once the paint is dry, glue a small picture of each planet onto the corresponding Styrofoam ball.

  4. Use string or yarn to hang the planets from the hanger in the correct order from the Sun. Make sure the distances between the planets are roughly proportional to their actual distances in the solar system.

  5. Hang your solar system mobile where you can admire your handiwork!

3. Create a Mini-Rocket

- **Materials:** Empty plastic bottle, cardboard, tape, baking soda, vinegar, cork (that fits snugly in the bottle opening), safety goggles
- **Instructions:**

  1. Decorate the bottle to look like a rocket using cardboard for fins and a nose cone. Use colorful markers and stickers to make it look realistic.

  2. Go outside to a clear area, away from buildings and trees. Wear safety goggles to protect your

eyes.

3. Pour some vinegar into the bottle (about 1/4 full).

4. Wrap a tablespoon of baking soda in a tissue and carefully drop it into the bottle.

5. Quickly put the cork in the bottle opening and stand the bottle upside down on a flat surface.

6. Stand back and watch your rocket blast off! (The reaction between baking soda and vinegar creates gas that builds up pressure and launches the bottle.)

7. Experiment with different amounts of baking soda and vinegar to see how it affects the rocket's launch height.

4. Design Your Own Alien

- **Materials:** Paper, crayons, markers, colored pencils, imagination!
- **Instructions:**

  1. Imagine what life might be like on another planet. What kind of environment would it have? What challenges would the creatures face?

  2. Use your art supplies to draw your own unique alien creature. Get creative with colors, shapes, and features. Think about how its environment might influence its appearance.

  3. Give your alien a name and write a short story about its life on its home planet. What does it eat? How does it move? Does it have any special abilities?

5. Make a Crater in a Box

Materials: A shallow box (like a shoe box), flour, cocoa powder, a small rock or ball, a ruler, a pencil
Instructions:

- Fill the box with a layer of flour, smoothing it out so it's level.

- Sprinkle a thin layer of cocoa powder on top to represent the surface of the moon or another planet.

- Drop the small rock or ball from a height to create a crater.

- Use a ruler to measure the depth and diameter of the crater.

- Experiment by dropping rocks from different heights or using different-sized rocks to see how the craters change.

6. Create a Comet with Dry Ice

Materials: Dry ice (handle with care!), a bowl of warm water, a spoon, food coloring, gloves, a towel, and a safety mask
Instructions:

- Place a small piece of dry ice in a bowl of warm water and watch it sublimate, creating a fog that looks like a comet's tail.

- Add food coloring to the water to make the tail of the comet more colorful and visually striking.

- Use a spoon to gently touch the dry ice, creating bursts of vapor.

- (Ensure kids are supervised with dry ice, as it can be dangerous if mishandled.)

7. Make a Star Chart

Materials: Blank paper, pen or pencil, star stickers, a reference star map (available online)
Instructions:

- Have the kids look at the night sky (or use an online star map) to find constellations that are visible in their location.

- Draw a chart of the night sky on the blank paper and mark where the stars and constellations are.

- Use star stickers to represent the stars.

- Compare their chart with the actual sky at night, seeing how many constellations they can find.

8. DIY Solar Oven

Materials: A pizza box, aluminum foil, black construction paper, plastic wrap, tape, and a thermometer
Instructions:

- Cut a flap in the top of the pizza box and line the inside with aluminum foil.

- Tape black construction paper to the bottom of the box to absorb heat.

- Tape plastic wrap over the opening to trap heat inside.

- Place a thermometer inside the box and check the temperature as the box "cooks" in the sunlight.

- Use the solar oven to melt chocolate or heat up s'mores ingredients, demonstrating how solar energy can be used to cook food!

9. Make Your Own Astronaut Helmet

Materials: Large clear plastic bowl, aluminum foil, cardboard, duct tape, scissors
Instructions:

- Use the plastic bowl as the base for the helmet.

- Cut a hole in the bottom big enough to fit your head through.

- Cover the outside of the bowl with aluminum foil to give it a shiny, metallic look.

- Cut out a face shield from clear plastic or acetate and tape it to the opening.

- Add cardboard and duct tape to create straps to secure the helmet.

- Put it on and pretend to be an astronaut exploring space!

10. Space-Themed Sensory Bottle

Materials: Clear plastic bottle, water, glitter, glow-in-the-dark stars, beads, small figurines, food coloring
Instructions:

- Fill the bottle with water, leaving a little space at the top.

- Add glitter, glow-in-the-dark stars, and beads to create a cosmic effect.

- Use food coloring to make the liquid darker (optional) to make it look more like deep space.

- Seal the bottle tightly with glue or tape to prevent spills.

- Shake the bottle to watch the "stars" swirl around, creating a calming space scene!

Space Quiz

Chapter 1: Introduction to Space

1. **What is the name of the galaxy we live in?**

   - o A) Andromeda
   - o B) Milky Way
   - o C) Whirlpool
   - o D) Triangulum

2. **Which of the following planets is known as the "Red Planet"?**

   - o A) Venus
   - o B) Mars
   - o C) Jupiter
   - o D) Saturn

3. **What is the primary element in the Sun's core?**

   - o A) Oxygen
   - o B) Hydrogen
   - o C) Carbon
   - o D) Helium

Chapter 2: Our Solar System

4. **Which planet is closest to the Sun?**

   o A) Mercury

   o B) Venus

   o C) Earth

   o D) Mars

5. **What are Saturn's rings made of?**

   o A) Water

   o B) Ice and rock particles

   o C) Gas

   o D) Dust

Chapter 3: Stars and Constellations

6. **What is the process that powers stars, including our Sun?**

   o A) Fusion

   o B) Fission

   o C) Combustion

   o D) Photosynthesis

7. **Which of the following is the brightest star in the night sky?**

   o A) Polaris

   o B) Sirius

- C) Betelgeuse

- D) Rigel

8. **Which constellation is known for its "belt" of three stars in a line?**

- A) Ursa Major

- B) Orion

- C) Leo

- D) Taurus

Chapter 4: The Moon and Its Phases

9. **How many phases does the Moon go through in a complete cycle?**

- A) 4

- B) 5

- C) 8

- D) 10

10. **What causes the phases of the Moon?**

- A) The Moon's orbit around Earth

- B) The Sun's distance from Earth

- C) The Earth's rotation

- D) The Moon's orbit around the Sun

Chapter 5: Space Exploration

11. What was the name of the first manned space mission to land on the Moon?

    o A) Apollo 11

    o B) Apollo 13

    o C) Soyuz 1

    o D) Gemini 4

12. Which space agency launched the first human to space?

    o A) NASA

    o B) Roscosmos

    o C) ESA

    o D) SpaceX

Chapter 6: Space Phenomena

13. What is a black hole?

    o A) A star that is cooling down

    o B) A region of space where gravity is so strong that nothing can escape

    o C) A large gas cloud in space

    o D) A dark spot on the Sun's surface

14. What is the name of the process by which stars produce energy?

    o A) Photosynthesis

    o B) Nuclear Fusion

- C) Chemical Reactions

- D) Gravitational Collapse

Chapter 7: Space Activities

15. What materials do you need to create a simple rocket using vinegar and baking soda?

- A) Plastic bottle, cardboard, tape, vinegar, baking soda

- B) Aluminum foil, plastic bag, water, glue

- C) Paper, straw, glue, sugar

- D) Cardboard box, rubber bands, water

16. What is the purpose of a constellation viewer?

- A) To measure the distance between stars

- B) To project constellations onto a wall or ceiling

- C) To capture images of distant planets

- D) To map the surface of the Moon

Chapter 8: Aliens and Extraterrestrial Life

17. What does SETI stand for?

- A) Search for Extraterrestrial Intelligence

- B) Solar Exploration Technology Initiative

- C) Space Exploration and Technology International

    o D) Solar Evolutionary Terrestrial Interactions

18. What is the closest star to Earth, after the Sun?

    o A) Alpha Centauri

    o B) Sirius

    o C) Proxima Centauri

    o D) Vega

Bonus Questions

19. What is the name of the largest planet in our solar system?

    o A) Saturn

    o B) Uranus

    o C) Jupiter

    o D) Neptune

20. Which spacecraft was the first to reach the outer planets, including Jupiter and Saturn?

    o A) Voyager 1

    o B) Apollo 11

    o C) Curiosity Rover

    o D) Hubble Space Telescope

Answers

1. B) Milky Way

2. B) Mars

3. B) Hydrogen

4. A) Mercury

5. B) Ice and rock particles

6. A) Fusion

7. B) Sirius

8. B) Orion

9. C) 8

10. A) The Moon's orbit around Earth

11. A) Apollo 11

12. B) Roscosmos

13. B) A region of space where gravity is so strong that nothing can escape

14. B) Nuclear Fusion

15. A) Plastic bottle, cardboard, tape, vinegar, baking soda

16. B) To project constellations onto a wall or ceiling

17. A) Search for Extraterrestrial Intelligence

18. C) Proxima Centauri

19. C) Jupiter

20. A) Voyager 1

Glossary of Space Terms

- **Asteroid:** A rocky object that orbits the Sun, smaller than a planet. Most asteroids are located in the asteroid belt between Mars and Jupiter.

- **Astronaut:** A person trained to travel and work in space. Astronauts undergo rigorous physical and mental training to prepare for the challenges of spaceflight.

- **Atmosphere:** The layer of gases that surrounds a planet. Earth's atmosphere protects us from harmful radiation and provides the air we breathe.

- **Black hole:** A region of spacetime where gravity is so strong that nothing, not even light, can escape. Black holes are formed from the collapse of massive stars.

- **Comet:** A ball of ice and dust that orbits the Sun, often with a long tail. Comets are sometimes called "dirty snowballs."

- **Constellation:** A group of stars that form a pattern in the sky. Constellations have been used for navigation and storytelling for centuries.

- **Exoplanet** – A planet that orbits a star outside of our solar system.

- **Galaxy:** A vast collection of stars, gas, dust, and planets. Our solar system is located in the Milky Way galaxy.

- **Gravity:** The force that attracts objects with mass towards each other. Gravity is what keeps us on Earth and what causes planets to orbit stars.

- **Hawking Radiation** – Theoretical radiation emitted by black holes, as proposed by physicist Stephen Hawking, which can cause black holes to lose mass over time.

- **Light-year:** The distance that light travels in one year. Light travels at a speed of about 186,000 miles per second.

- **Microgravity:** A condition of very weak gravity, like what astronauts experience in space. Microgravity can cause changes in the human body, such as bone loss and muscle weakness.

- **Nebula:** A cloud of gas and dust in space. Nebulae are often the birthplaces of stars.

- **Orbit:** The path that an object takes around another object in space. Planets orbit stars, and moons orbit planets.

- **Planet:** A large celestial body that orbits a star. Planets are generally spherical in shape and have cleared their orbits of other debris.

- **Quasar** – Extremely bright and energetic objects powered by supermassive black holes in the centers of galaxies.

- **Rocket:** A vehicle that uses the expulsion of hot gas to propel itself. Rockets are used to launch spacecraft into space.

- **SETI** – Search for Extraterrestrial Intelligence, focusing on efforts to detect alien life through radio signals and other forms of communication.

- **Satellite:** An object that orbits a planet. Satellites can be natural, like the Moon, or artificial, like the International Space Station.

- **Solar system:** A star and all the objects that orbit it. Our solar system includes the Sun, eight planets, dwarf planets, asteroids, and comets.

- **Space probe:** An unmanned spacecraft sent to explore planets, moons, and other celestial bodies. Space probes carry scientific instruments to collect data and send it back to Earth.

- **Star:** A giant ball of hot gas that produces light and heat. Stars are powered by nuclear fusion, the process of combining lighter elements into heavier ones.

- **Telescope:** An instrument used to view distant objects in space. Telescopes can be ground-based or space-based.

- **UFO/UAP** – Unidentified Flying Objects/Unidentified Aerial Phenomena, referring to unexplained sightings in the sky, some of which are speculated to be alien technology.

- **Universe:** All of space and everything in it. The universe is vast and contains billions of galaxies.

- **Wormhole** – A hypothetical tunnel in space-time that could provide shortcuts for interstellar travel.

As you've journeyed through the pages of this book, you've learned about the incredible vastness of space, the wonders of our solar system, and the mysteries that lie beyond. But most importantly, we hope you've gained a deeper appreciation for our place in the universe and the interconnectedness of all things. Remember, we are all made of stardust, and the cosmos is within each of us.

Review Request

Review us on **Amazon US!**

or any Amazon site where you purchased this book!

Check my other ongoing books in the series!

Scan the QR!

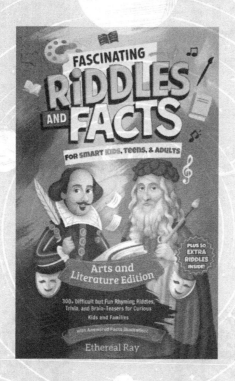

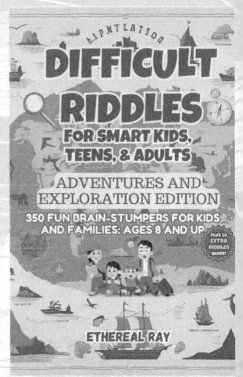

Made in the USA
Coppell, TX
15 December 2024

42608322R00059